Exploring Saturn

Dan Bortolotti

FIREFLY BOOKS

A FIREFLY BOOK

Published by Firefly Books Ltd. 2003

First printing

Publisher Cataloging-in-Publication Data (U.S.)
(Library of Congress Standards)

Bortolotti, Dan.
    Exploring saturn / Dan Bortolotti.—1st ed.
[64] p. : col. ill., photos. ; cm.
Summary: History of the exploration of Saturn from ancient astronomers to latest NASA findings. Includes website resources, and location guide for backyard observations.
ISBN 1-55297-766-8
ISBN 1-55297-765-X
1. Saturn (Planet)—Juvenile literature. (1. Saturn (Planet)). I. Title.
523.46 21   QB671.B67 2003

National Library of Canada Cataloguing in Publication Data

Bortolotti, Dan
    Exploring Saturn / Dan Bortolotti.
ISBN 1-55297-766-8 (bound).—ISBN 1-55297-765-X (pbk.)
    1. Saturn (Planet)—Exploration—Juvenile literature. I. Title.
QB671.B66 2003   j523.46   C2003-902811-9

Published in the United States in 2003 by
Firefly Books (U.S.) Inc.
P.O. Box 1338, Ellicott Station
Buffalo, New York 14205

Published in Canada in 2003 by
Firefly Books Ltd.
3680 Victoria Park Avenue
Toronto, Ontario, M2H 3K1

Design: Gillian Stead and Tinge Design Studio

Printed in Canada by Friesens, Altona, Manitoba

The Publisher acknowledges the financial support of the Government of Canada through the Book Publishing Industry Development Program for its publishing activities.

Previous page:
Serene and symmetrical, Saturn is the showpiece of the solar system. This Hubble Space Telescope shot was taken in November 2000. *Courtesy Space Telescope Science Institute.*

Photo Credits

All images courtesy of NASA/JPL/Caltech except:

p.8 John Sanford/Science Photo Library

p.10 Science Photo Library

p.11 Science, Industry & Business Library/New York Public Library/Science Photo Library

p.13 Science Photo Library, D.A. Calvert, Royal Greenwich Observatory/Science Photo Library (telescope)

p.14 Science Photo Library

p.16 Claus Lunau/FOCI/Bonnier Publications/Science Photo Library

p.30 Peter Ryan/Science Photo Library

p.32 NASA/Science Photo Library

p.33 Seth Shostak/Science Photo Library

p.38 Julian Baum/Science Photo Library

p.43 S. Sheppard/D. Jewitt/University of Hawaii

p.48 Joe Skipper/Reuters

p.49 Detlev Van Ravenswaay/Science Photo Library

p.51 Science Photo Library

p.54 Camay Sungu/Reuters

p.57 Courtesy of Starry Night, a division of Space Holding, Corp.

Illustrations

p.22/23 Christine Gilham

p.46 George Walker

Table of Contents

✧

The Distant Wanderer

No one knows who spotted it first. Perhaps it was a prehistoric hunter on his way back to the cave with an armload of mammoth meat. He may have glanced up at the sky and noticed a yellowish point of light – brighter than most, but otherwise no different from the countless stars he saw each night. Not worth pausing to stare at, especially if it would make him late for dinner.

This false-color image of Saturn was taken by the Hubble Space Telescope on January 4, 1998.

Fast-forward thousands of years. Astrologer-priests in China and the Middle East have noticed that this yellow star moves, and they try to predict where and when it will show up in the future. These early skywatchers were the first people to study the planet we now call Saturn, the sixth from the sun and the most distant one visible with the naked eye.

In the centuries since, humans have slowly pieced together Saturn's puzzles. We've built bigger and better telescopes to study its face and its rings.

We've made educated guesses about what's beneath its veil of clouds. We've loaded scientific gizmos onto rockets and fired them into the dark freezer of space to be our eyes and ears. The most recent of these spacecraft, *Cassini*, will explore Saturn from 2004 to 2008. It will answer many of the questions scientists have about the planet – and will raise many more.

No astronaut will ever walk on Saturn, and no one will gape at it through the window of a spaceship any time soon. Most people have never even seen it through a telescope. The planet hangs prominently in the night sky for several months every year, yet few of us ever notice it. Still, after the Sun and the Moon, Saturn is the most instantly recognized object in the solar system. When children scrawl their first space pictures, one of the planets always has a ring around it.

Saturn may be a cold, faraway ball of gas, but its beauty has made it a unique part of our world.

◁ Cassini *sent home its first photo of Saturn on October 21, 2002, five years after its launch and still 285 million km (177 million miles) from the planet. Titan, Saturn's largest moon, is at the upper left.*

Old Man In The Sky

Our ancestors, without the benefit of digital watches, global positioning systems or The Weather Channel, paid more attention to the sky than we do. By careful observation, they noticed that five "stars" changed in brightness and position from week to week, and that they followed a similar path to that of the sun and moon – a narrow band of sky called the zodiac.

The Greeks called these movable stars *planetes*, which means "wanderers," and named them after gods and deities. The slowest they named for Kronos, who was Greek mythology's lousiest father – fearing his children would steal his throne, he ate them. The ancient Romans later associated the hungry Kronos with their god of agriculture and called the planet Saturn. Each December, the Romans held a feast in his honor called Saturnalia, and some of its traditions live on in modern Christmas celebrations.

In the Middle Ages, people associated Saturn with the Grim Reaper or Father Time, calling him "the bringer of old age."

Today, the word *saturnine* means gloomy, sullen, or sluggish – perhaps an allusion to the slowest of the wandering stars. The planet's name also gives us the word *Saturday* – the seventh day of the week, named for the most distant of the seven objects in the solar system known since ancient times.

Many cultures believed that the planets' positions have an effect on human affairs. More than 2,600 years ago an astrologer in Babylonia was concerned when Saturn appeared close to the moon: "When a planet changes color opposite the Moon … lions will die … cattle will be slain."

By the time they were conquered by the Persians in 539 BC, the Babylonians had become remarkably skilled at measuring and forecasting celestial events. Yet they made no attempt to come up with a model that explained *why* the sun, moon and planets move they way they do. For that, we have to look to the Greeks.

◁ *Mars, Jupiter and Saturn all appear in this photograph of the zodiac constellation Leo. While the patterns of stars do not appear to change, the positions of the planets can be easily followed over the course of days and weeks.*

Revolutionary Ideas

The Greeks were the first to come up with a scientific model to explain the motions of the planets. In the fourth century BC, the unpronounceable Eudoxus of Cnidus imagined that the Sun, Moon, stars and planets were carried around the Earth by a series of 27 spheres – 4 of them for Saturn alone.

A few decades later, Aristotle built upon this idea, reasoning that the planets must travel around the Earth in perfect circles. But this idea had more to do with philosophy than actual astronomy, since it did not explain the planets' two most puzzling behaviors: First, unlike the unwavering stars, the planets appear brighter on some nights than others. And second, though they usually move from west to east against the background of stars, they sometimes drift backwards – astronomers call this "retrograde motion." If the planets orbited the Earth in circles, neither of these behaviors would make any sense.

In the fourth century BC, Greek philosophers such as Aristotle tried to explain how Saturn and the other planets moved through the sky.

Some 500 years after Aristotle, Ptolemy tried to improve the Greek models. He suggested that each planet moves in a circle, and that these circles move in bigger circles around the Earth. It was a wickedly complicated model, but it did explain the changing brightness of the planets, and it predicted their positions quite reliably. (And when it didn't, he just added more circles.) For example, in Ptolemy's model the retrograde motion of Saturn lasted a maximum of 140 days, eight hours. He was off by less than a day.

Ptolemy's system endured for 14 centuries, until Nicolaus Copernicus made it his mission to dismantle it. Copernicus hung on to the incorrect idea of circles within circles – his model had 34 in total, including 5 for Saturn. But he added a revolutionary twist, suggesting that the planets revolve around the Sun, not the Earth. Copernicus didn't live to see his ideas accepted. When he died in 1543, the same year his work was published, his model not only offended most Europeans' religious sensibilities, but it was so difficult to understand that it was largely rejected.

It fell to the young German astronomer Johannes Kepler to complete the model. Kepler hit on two radical ideas: First, the planets moved around the Sun in ellipses – flattened circles, not perfectly round ones. He also discovered that the closer a planet gets to the Sun during its orbit, the faster it travels.

Just like that, Kepler answered the two toughest questions about planetary motion: The elliptical orbits explained why the distance between the planets and the Earth changed, and therefore why their brightness varied. As for Saturn and the other outer planets changing direction, this is simply an illusion caused by the Earth speeding past these distant worlds. Imagine one racing car passing another – to the driver of the faster car, the slower vehicle appears to move backwards as it is overtaken.

Kepler published these discoveries in 1609, the same year an Italian scientist arrived in Padua with a new invention that was about to change the way the world looked at Saturn.

For 1,400 years, scientists and philosophers followed Ptolemy's model, in which the Sun, Moon, planets and stars traveled around the Earth.

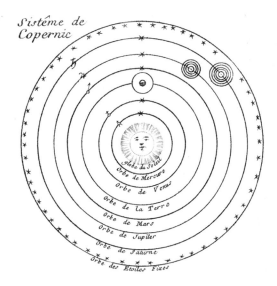

Copernicus showed that the planets revolve around the Sun, not the Earth, though few people paid attention to his idea when it was published.

"A Most Extraordinary Marvel"

In the autumn of 1609, Galileo Galilei was spending more and more time tinkering in his workshop in Padua, Italy. Earlier in the year, he had been introduced to an invention that was spreading swiftly across Europe: the telescope. Up to now it was mainly used to scan the horizon for distant ships. But Galileo had other ideas for the telescope – he would use it to bring the heavens to Earth.

Born in Pisa in 1534, Galileo was a brilliant scientist and, from 1609 until the 1630s, he was also the most skilled telescope-maker in the world. Before Galileo, telescopes could magnify objects no more than 3 times. But after only a few months of fiddling with its design, Galileo created a telescope that magnified objects 20 to 30 times.

Even today people gasp when they see Saturn and its rings through a telescope for the first time.

This sketch, made by Galileo in 1616, clearly shows how Saturn's ring appeared as "handles" in his early telescope. The mysterious shape baffled astronomers for almost 60 years.

One can only guess what sounds sputtered from Galileo's lips when, on July 25, 1610, he aimed his scope toward the planet and squinted at the eyepiece. This is how he explained it in a letter five days later: "I have discovered a most extraordinary marvel … the planet Saturn is not one alone, but is composed of three, which almost touch one another. They are completely immobile and are situated in this manner: oOo."

Galileo had just become the first human to see Saturn's rings, though his telescope was not powerful enough for him to recognize what they were. He thought he was seeing three separate objects. The previous January, Galileo had discovered Jupiter's four large moons. But if these two companions were Saturn's moons, they were enormous. And besides, they didn't move each night like Jupiter's satellites did. Frustrated and confused, Galileo temporarily lost interest in observing Saturn.

He tried again in December 1612 and was amazed to find that the planet's sidekicks had disappeared. "Has Kronos perhaps devoured his own children?" Galileo quipped. By 1616, the planet looked different once again – rather than being "composed of three," it appeared to have what Galileo called *ansae*, Latin for "handles." What was going on?

Galileo was never able to explain what caused Saturn's unusual appearance. He died in 1642, and it was to be more than a decade before the mystery was solved.

Galileo demonstrates his telescope in Venice in 1609. Even the scientist himself first thought of maritime or military uses for the new invention.

Eye Spy

While no one knows for sure who built the first spyglass – the predecessor of the astronomical telescope – the credit usually goes to a Dutch optician named Hans Lippershey. As the story goes, in 1608 some playful children in his shop noticed that, if they looked through one lens placed in front of another, the steeple of the town's church appeared closer. The inspired Lippershey later fitted the two lenses into a tube to create the first telescope. He was unable to patent the invention, however, and by the following year it was showing up in England, France and Italy.

Galileo's early telescopes had lenses less than an inch (2.5 cm) in diameter and magnified objects 20 to 30 times.

Handles Messiah

After Galileo's discovery in 1610, astronomers routinely observed Saturn and sketched its changing shape. But it was not until the 1650s that any of them made progress in explaining the "handles" and the planet's varying appearance. The challenge was no longer building bigger telescopes – by the mid-17th century the best instruments could magnify the planet 50 times, more than enough to see the ring structure. The problem was understanding what they saw.

One scientist proposed that the *ansae* were two crescent-shaped structures attached to the planet. Another thought Saturn was shaped like an egg, with four dark patches on it. Still others suggested the planet emitted some sort of vapor around its equator.

The man who would eventually solve the

Christiaan Huygens was not only an astronomer, but a brilliant physicist who pioneered the study of light waves and built the most accurate clock of his era.

mystery was Christiaan Huygens, a Dutch physicist and astronomer. On March 25, 1655, seated behind a 3.7-meter (12-foot) telescope he had built himself, Huygens noticed what appeared to be a tiny star to the west of Saturn. When he looked again the next night, the object had moved, suggesting that it wasn't a star at all. Huygens had discovered Saturn's largest moon, later to be named Titan. He continued studying the planet through his eyepiece and turned his mind to its greatest mystery: What could explain the changing *ansae*?

By 1656, his careful observations made it clear that only one model made sense. Saturn must be "surrounded by a thin, flat ring, nowhere touching" the planet's surface, and this ring is tilted with respect to Saturn's orbit – the same way the Earth is tilted on its axis. This theory explained why, over the course of Saturn's 30-year orbit, an observer on Earth would sometimes see the rings face-on from above or below (when they would be most obvious) and at other times edge-on (when they would be virtually invisible).

Now it was clear why Galileo first saw "three stars" in his telescope – he was actually seeing the ring plane at a sharp angle. When he joked that "Kronos devoured his children" two years later, he was looking at the rings edge-on. And when the ring plane was tipped toward him in 1616, the planet appeared to have handles.

It took Huygens three years to write a full account of this theory, but he finally published it in 1659. He was initially criticized, but by 1671 the grumbling had stopped and astronomers accepted his explanation of Saturn's unique shape.

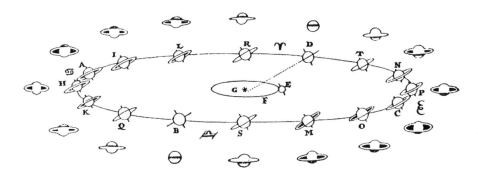

Published in 1659, Huygens's study of Saturn finally explained how the tilt of its ring was responsible for the planet's changing appearance.

Did the Ancients Know About the Rings?

Although modern science didn't identify Saturn's rings until the 1600s, some people believe that ancient cultures knew about them centuries before.

In Classical literature and art, the god Kronos (the Greek name for Saturn) is occasionally described as being in chains, veiled or encircled by a snake. Could this have been a poetic way of describing a planet with a ring, as some have suggested? That's a bit of a stretch.

Others point to the Dogon tribe of Mali, West Africa. French anthropologists who lived among them in the mid-20th century claimed their mythology has described Saturn as a ringed planet for thousands of years. Most experts now believe that this was a misunderstanding.

In New Zealand, the Maori have long referred to Saturn (or Jupiter in some accounts) as Pareärau, an ancient name that means "surrounded by a headband." Because it is traditional for Maori women to wear a headband of greenery as a symbol of mourning, Pareärau was also called the Widow Star.

A Planet Is Born

The story of Saturn begins 4.5 billion years ago. Back then, the newly formed Sun was at the center of a huge, swirling disk of gas and dust. Over a period of about 100,000 years, this dust accumulated into small solid bodies which collided and combined to form the rocky cores of the planets. Particles of dust and ice were more plentiful in the outer solar system, where the heat of the Sun was weaker. Here the planetary cores – on their way to becoming Jupiter, Saturn, Uranus and Neptune – reached 10 to 15 times the mass of Earth. Their strong gravity attracted the surrounding hydrogen and helium gas, and over the next 100,000 years the infant planets doubled in mass. The giant planets collected more and more gases for another several million years.

Saturn and the other planets were formed from the dust and gas that swirled around the newly formed Sun more than 4.5 billion years ago.

Saturn today is a giant globe of gas spanning more than 120,000 km (74,500 miles) at its equator – almost 10 times the diameter of Earth. If it were hollow, more than 760 Earths would fit inside it, although it is only about 95 times heavier than our home planet. In fact, relative to its size, Saturn is the lightest planet in the solar system: Earth's average density is about 5.5 times that of water, while Saturn is only 70 percent as dense as water. In theory, Saturn could float.

If we could slice the planet in half, the inside would appear to have three main regions. At the center is a solid core of rock, probably about the size of Earth, but far more dense. Wrapped around this core is a layer of liquid metallic hydrogen. The planet's outer layer, about 30,000 km (18,600 miles) thick, is molecular hydrogen (H_2) in liquid form, gradually changing to a gas as it moves toward the surface.

Unlike on Earth, where it's clear where the land and sea stop and the atmosphere begins, this boundary is blurry on Saturn. There is no single point at which the liquid hydrogen becomes a gas. Instead, there is a transitional area where the hydrogen is in what's called a supercritical state, meaning that it has some properties of a liquid and some of a gas. On the whole, the atmosphere is more than 93 percent hydrogen and 6 percent helium, with tiny amounts of other gases.

Saturn has three vertical cloud layers, made up of frozen particles suspended in gas. The innermost clouds are made of water, above which is a layer of ammonium hydrosulfide clouds and, on top of that, a layer of ammonia clouds. These swirl around the planet at breakneck speed, organizing themselves into horizontal bands that make the surface of Saturn appear striped, although this is not easily visible from Earth. The winds around the equator move faster than those near the poles, topping out at 1,600 km (1,000 miles) per hour. By comparison, the most devastating tornadoes on Earth produce wind speeds of around 400 km (250 miles) per hour.

Short-lived storms occasionally appear, and once during every orbit Saturn is hit by a huge tempest that lasts about a month. Known as the Great White Spot, this swirling storm was first

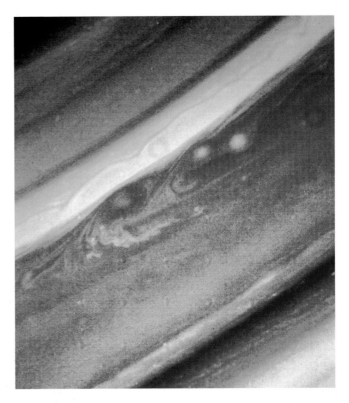

Horizontal bands of cloud have been colored to make them easy to see in this image of Saturn's surface. The light blue ribbon above the three ovals is traveling about 500 km (300 miles) per hour.

spotted in 1876 and has appeared about every 29 or 30 years since, most recently in 1990. Astronomers believe the storm is caused when the atmosphere warms up at the end of the planet's summer, creating ammonia crystals that are then whipped around by extraordinarily strong winds.

Saturn is bitterly cold – its average temperature is around -185°C (-300°F). However, the planet actually releases more heat than it absorbs from the Sun. No one is quite sure why, but scientists have a

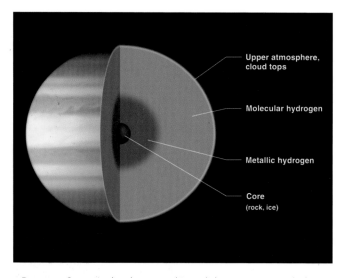

Upper atmosphere, cloud tops

Molecular hydrogen

Metallic hydrogen

Core
(rock, ice)

Between Saturn's cloud tops and its solid core, scientists believe there are two layers of hydrogen — one made up of gas and liquid, the other a highly compressed metallic liquid.

Saturn's Great White Spot is a storm that occurs near the equator about every 30 years. Smaller storms are also visible from time to time.

couple of ideas. It's possible that Saturn's huge envelope of gas has trapped much of the heat left over from when the planet was formed, and this heat is slowly escaping. The most popular theory, though, is that the heat is caused by friction as liquid helium sinks toward the core through the layers of lighter hydrogen.

Saturn has a faster rotation period than any planet but Jupiter. At the equator, the atmospheric currents make a complete circuit about once every 10 hours and 14 minutes; at the poles, a complete spin takes a half-hour longer. This causes Saturn to bulge noticeably, as though someone with a thumb and forefinger on the poles were squeezing it. All rotating bodies do this to some extent, but on the ringed planet it's quite pronounced. Earth is only about 42 km (26 miles, or 0.36 percent) wider at the equator, whereas Saturn's love handles make it 12,000 km (7,450 miles, or 10.9 percent) bigger around the middle.

At Saturn's interior, the solid core and liquid metallic hydrogen are also spinning, and this motion gives the planet its magnetic field, which is about 1,000 times stronger than Earth's. Our planet's magnetic field is tilted more than 11 degrees relative to its rotational axis, which is why compasses don't point directly at the north pole. Saturn's magnetic and rotational axes, however, are almost perfectly in line, and the poles are opposite to those on Earth. So a compass on Saturn would point almost due south.

Saturn's magnetic field reacts with charged particles emitted by the Sun to form glowing auroras around the poles, just like the northern and southern lights on Earth. These charged particles, called the solar wind, are steered around the planet by the magnetic field, creating a kind of bubble called a magnetosphere. The bubble is not spherical, though – on the side facing the Sun the magnetosphere is rounded, but on the night side of the planet it tapers off to form a long tail. Like a boat traveling through water, Saturn leaves a wake as the solar wind rushes past it. The enormous magnetosphere not only includes the planet and its ring system but it extends past the orbit of Titan, some 1.2 million km (750,000 miles) away.

In the early 1980s, the *Voyager* spacecraft discovered that Saturn sends out radio waves from within its core. By watching the patterns and changes in these radio waves, scientists can estimate how fast that core is rotating. Until then, the planet's rotation speed was measured by observing the clouds, which move at different speeds. They learned that a complete rotation of the core takes 10.66 hours.

Saturn's magnetic field interacts with charged particles from the Sun to create auroras, similar to the northern and southern lights on Earth.

Saturn orbits the sun at an average distance of about 1.43 billion km (888 million miles). As Kepler discovered in the 17th century, planets move more slowly as they get farther from the Sun, so Saturn's average velocity of 9.64 km (6 miles) per second seems plodding compared with Earth's rate of almost 30 km (18.5 miles) per second.

How Do You Figure That?

Long before the first spacecraft flew by Saturn in 1979, astronomers had a pretty good idea about the planet's distance, its mass and the contents of its atmosphere. They even suspected what its interior was made of. How did they figure all this out from more than a billion kilometers away?

Until quite recently, measurements of the solar system were relative, not absolute. Scientists were able to determine distances and sizes only in relation to one another – they knew Saturn is almost twice as far as Jupiter, for example, and that it's 95 times heavier than Earth. But they could not measure the planet's actual distance in kilometers or its mass in kilograms. Even today, the yardstick of the solar system is the astronomical unit (AU), or the average distance from the Sun to the Earth.

As soon as Copernicus abandoned the idea of an Earth-centered solar system in the 15th century, he was able to roughly work out the relative distances for each planet. By sketching a triangle with the Earth, Sun and a planet at the three points, he used geometry to determine their distances in AU. Of course, because he was unaware the planets had elliptical orbits, his numbers were not precise – he reckoned the average distance to Saturn was about 9 AU – though he was off by only 6 percent.

In the early 1600s, Kepler showed that the planets actually orbit in ellipses. He also discovered that their distances were related to the time it takes to complete their orbits. When you plug Saturn's orbital period (29.46 years) into Kepler's formula – his Third Law of Planetary Motion, published in 1619 – you get its average distance from the sun: 9.55 AU. This is the same figure we accept today.

These relative measurements weren't completely satisfying, but scientists knew that when they found one absolute value, the others would fall into place. So they set out to get that one reliable number.

Hold your thumb at arm's length and look at it first with one eye, then the other. You'll notice that your thumb appears to move relative to the background. This is called parallax – the apparent change in the position of an object when viewed from different places. In October 1672, French astronomers Abbé Jean Picard and Jean Richer used this ingenious technique to measure the distance between Earth and Mars. Richer hopped a ship for South America while Picard stayed in Paris. The two men took simultaneous observations of Mars from these two distant locations (the global equivalent of your left and right eyes) and measured the angle between their lines of sight. Using geometry they then calculated the distance to Mars at about 52.5 million km (32.6 million miles).

Now Picard and Richer had the benchmark needed to figure out other distances, and they determined the Earth-Sun distance (the AU) to be about 140 million km (87 million miles) – just 7 percent off the actual figure. Saturn, at 9.55 AU, would therefore be about 1.34 billion km (832 million miles) away; the actual figure is 1.42 billion km (882 million miles). In the 1760s, scientists improved these numbers by observing Venus from several locations around the world using a similar method. Two centuries later, radar measured the AU much more precisely, but parallax is still an accurate technique – spacecraft even use it to determine the distance to stars hundreds of light years away.

Determining the mass of Saturn – relative to that of the Sun – became possible in 1687, when Isaac Newton published his law of universal

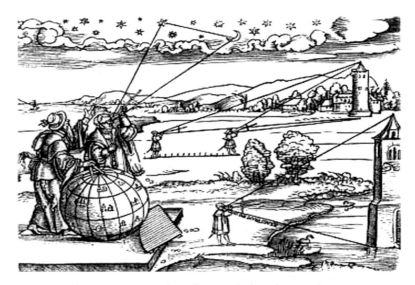

Early astronomers used parallax to calculate distances between objects with remarkable accuracy.

gravitation. This law explained that all objects exert a force of attraction, and the strength of that force depends on the mass of the objects and the distance between them.

Using a modified version of Kepler's third law, Newton made it possible to determine the mass of the Sun – again, all you needed to know was the distance at which a planet orbited it, and the duration of that orbit. In the same way, you could also determine the relative mass of a planet by observing the motion of one of its satellites. Knowing that Titan orbits Saturn in about 16 days at an average distance of 1.2 million km (745,000 miles), astronomers could now calculate that the mass of Saturn was about 3,500 times less than that of the Sun.

Even before the space age, scientists learned to identify different gases in Saturn's atmosphere by measuring the unique way each absorbs light.

Scientists can also get an accurate picture of Saturn's atmosphere without sampling it directly. They had long assumed that hydrogen and helium must make up the vast majority of it – these are the most common materials in the solar system, and their low mass fits with Saturn's low density. But what about the tiny amounts of other gases?

The secret lies in the different ways gases absorb light. Anyone who has studied a rainbow knows that white light can be broken into a spectrum of colors, with red at one end and blue at the other. Every substance absorbs these colors in a different way. Methane (CH_4), for example, absorbs much of the light at the red end of the spectrum and reflects much of the blue. As early as the 1860s, astronomers were using spectroscopes – which measure these different bands of light – to identify gases by the unique way each absorbs sunlight. By the 1930s they had discovered methane in Saturn's atmosphere, and as spectroscopes improved they found tiny amounts of many other gases.

From the outside of the planet, scientists moved to Saturn's interior. How exactly does one determine what's beneath thousands of kilometers of gas and liquid? The answer emerged slowly. Once the volume and mass of Saturn were known, it was a

To convert these relative masses to kilograms, however, scientists must use a value called the gravitational constant, affectionately known as Big G. It was not until 1798 that English physicist Henry Cavendish determined its approximate value. However, it's so difficult to measure that even today Big G remains one of the most elusive numbers in physics.

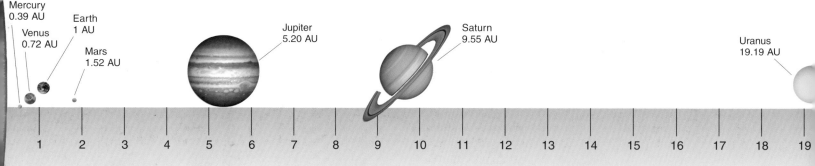

Mercury
0.39 AU

Venus
0.72 AU

Earth
1 AU

Mars
1.52 AU

Jupiter
5.20 AU

Saturn
9.55 AU

Uranus
19.19 AU

1 2 3 4 5 6 7 8 9 10 11 12 13 14 15 16 17 18 19

simple matter to figure out that its density is only about one-eighth that of Earth. Under the clouds, then, Saturn cannot be a huge ball of iron or rock – or even water, ammonia or methane. If it were, it would be far heavier than we know it to be.

Astronomers also considered how the mass of the planet was distributed – it clearly wasn't a perfect sphere. The small, slow changes in the motion of Saturn's nearby moons, which had been observed for centuries, indicated that the planet must be millions of times more dense at the center than at the surface.

The next clue came from the planet's magnetic field. While it took modern spacecraft to confirm that Saturn had one (it's too weak to detect from Earth), astronomers had predicted it long before. A planet's magnetic field is generated by the rotation of a material that conducts electricity. On Earth it originates in the iron core, but astronomers were certain there wasn't enough iron in the outer solar system to generate a magnetic field around Saturn. There was certainly lots of hydrogen, but this element is too light to account for the density of the interior, and besides, it doesn't conduct electricity.

Or maybe it does – if the conditions are right. In the 1930s, physicists suggested that if hydrogen

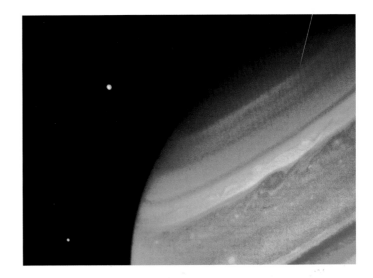

Thanks to Isaac Newton, astronomers were able calculate the relative mass of Saturn by measuring the distance and orbital periods of its moons.

were subjected to enormous pressure – like the pressure near the center of the gas planets – it would behave like a metal. This substance, known as liquid metallic hydrogen, fits with everything we know about Saturn's interior: It's extremely dense, it conducts electricity and it's made from the most common of elements.

Discoveries like this may give the impression that scientists know Saturn inside and out. But the ringed world still holds many mysteries, including the nature of its most famous feature: the rings.

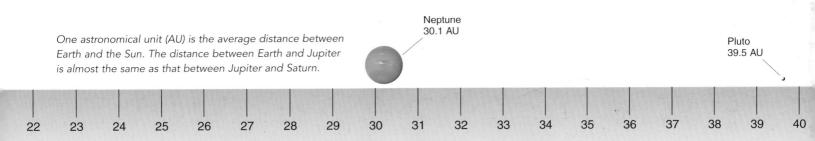

One astronomical unit (AU) is the average distance between Earth and the Sun. The distance between Earth and Jupiter is almost the same as that between Jupiter and Saturn.

Neptune
30.1 AU

Pluto
39.5 AU

22   23   24   25   26   27   28   29   30   31   32   33   34   35   36   37   38   39   40

Cassini and the Ring Masters

Saturn's bland, butterscotch globe isn't terribly alluring in a small telescope. It's the magnificent rings that steal the show.

The rings of Saturn are one of the solar system's most beautiful sights and most enduring mysteries. Every new discovery, from Christiaan Huygens to *Voyager 2*, has raised more questions than it has answered.

When Huygens published his ring theory in 1659, he didn't get everything right. For one thing, he insisted the ring was solid, thousands of miles thick, an idea he held on to until he died. If it was invisible when viewed edge-on, he argued, it was because the edge did not reflect sunlight. Some of his colleagues disagreed, arguing that the ring was most likely made up of small particles.

For two centuries, astronomers debated whether Saturn's rings were solid or made up of individual particles. This color-enhanced image shows that the system is actually countless small "ringlets."

The debate continued for more than 200 years, with equally brilliant minds taking opposite sides. Giovanni Domenico Cassini, head of the newly opened Paris Observatory, didn't buy the idea of a solid ring. In 1675, he discovered there were actually two rings, separated by a gap that was later named the Cassini division. It seemed unlikely that two solid disks would circle around the planet. Yet more than a century later, William Herschel, one of the greatest observers of all time – he discovered Uranus in 1781 and two moons of Saturn eight years later – was still supporting Huygens's theory.

The mystery was not solved until the late 19th century. Astronomers discovered a dim third ring (later called the C ring) close to the planet's surface in 1850 and noticed that it was semi-transparent. One observer at the time wrote that "the planet is seen through it as through a film of smoke." Findings such as this prompted the University of Cambridge in England to offer a prize in 1857 to the scientist with the best explanation of the rings' structure. The winner was James Clerk Maxwell, who showed that a solid ring was impossible – Saturn's gravity would break it up. "The only system of rings which can exist," he wrote, "is one composed of an indefinite number of unconnected particles."

Maxwell, following Kepler's laws, also suggested that the particles must orbit Saturn at different speeds, with those closest to the planet traveling fastest. If the rotating rings were solid, the reverse would be true – the outer edge would be moving more quickly than the inner edge, as it does on a bicycle wheel. In 1895, American astronomer James Keeler used a spectroscope to measure the light reflected by the rings and discovered that the innermost particles were indeed the speediest. The solid ring theory had been dealt a fatal blow.

Voyager 1 *flew by the unlit side of the rings to get a perspective never visible from Earth. The darkest area is the dense B ring, while the bright narrow band outside it is the Cassini division.*

Saturn From A to G

Saturn's ring system is incredibly complex, and scientists are still trying to answer some basic questions about it – such as how and why it was formed in the first place. They do know that the trillions of particles are composed almost entirely of water ice, with traces of other substances. The overwhelming majority range from tiny specks to several meters across.

The fact that Saturn's rings lie along the same plane as its equator is no coincidence. The gravity is strongest there, and over millions of years the particles have fallen into line as they've marched around the planet. The ring system spans hundreds of thousands of kilometers, if you include the faint outer strands that are barely visible. By contrast, it's extremely thin – typically not more than 200 meters (220 yards), and in many places substantially less.

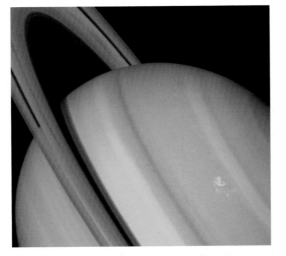

A good backyard telescope can often show the shadow that Saturn's rings cast on the planet's equator.

Unlike Saturn's satellites, which have colorful names borrowed from mythology, the seven distinct rings have simply been assigned a letter, from A to G in order of their discovery. Each is made up of thousands of smaller ringlets, most of which are roughly circular, though some are quite elliptical, while others are only arc segments, or incomplete circles. Only the two brightest – the A and B rings, separated by the Cassini division – are easily seen in a backyard telescope.

Scientists are still arguing about how the rings were formed. In 1848, French mathematician Edouard Roche came up with a suggestion that was largely ignored at the time but has since become

◁ *An artist's idea of what Saturn's rings might look like from up close. The vast majority of ring particles are probably tiny, with the largest about the size of a house.*

influential. When a large moon orbits close to a planet, the gravitational pull is stronger on the near side of the satellite than it is on the far side. That is why we have ocean tides, and why our moon bulges slightly toward the Earth. Roche calculated that if a moon got closer than 2.44 times the radius of Saturn (a distance now called the Roche limit) the tidal forces could tear it to pieces, leaving a ring of debris circling the planet.

Many modern scientists believe that the rings are indeed the remnants of one or more doomed satellites or comets, perhaps only tens of millions of years old. Others hold that the ring particles are leftovers from the formation of the planet 4.5 billion years ago – because they are so close to the planet, gravity did not allow them to coalesce and form a moon.

However they formed, the rings are certainly affected not only by the gravity of Saturn but by that of its moons, sometimes dramatically. In 1867, American astronomer Daniel Kirkwood noticed a relationship between the inner rings and the moon Mimas. This satellite orbits Saturn in about 22.6 hours, and Kirkwood pointed out that any ring particles orbiting in exactly half that time – at what's called a resonance point – would be regularly tugged by Mimas's gravity. His calculations revealed that particles moving at 11.3 hours would fall within the Cassini division. This may explain why the gap is there – particles originally occupying that space were pulled into a different orbit.

After all the breakthroughs of the late 19th century, little else was learned about Saturn's rings until an entirely new kind of scientist entered the picture in the 1970s. Its name was *Pioneer 11*.

In the 19th century, astronomer Daniel Kirkwood suggested that the gravity of Saturn's moon Mimas might be responsible for clearing the particles in Cassini division. In the image below, the division is the dark band in the rings; it is also visible in the bottom right corner of the photo at top.

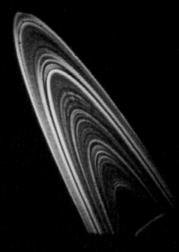

SPACECRAFT	LAUNCH	
PIONEER VENUS - MULTI-PROBE	08-08-78	
PIONEER VENUS - ORBITER	05-20-78	VENUS
VOYAGER 1	09-05-77	JUPITER
		SATURN
VOYAGER 2	08-20-77	JUPITER
		SATURN
		URANUS
VIKING 1	08-20-75	MARS
VIKING 2	09-09-75	MARS
HELIOS 1	12-10-74	SUN
HELIOS 2	01-15-76	SUN
PIONEER 10	03-10-72	
PIONEER 11	04-05-73	

Stepping into the Ring

Just as Galileo's telescope changed virtually everything we knew about Saturn, the images beamed back in 1979 from the *Pioneer 11* spacecraft, and later from *Voyager 1* and *Voyager 2*, forced scientists to reconsider three centuries of observation.

Pioneer 11 was launched on April 5, 1973, and arrived at Jupiter in December 1974. Its older brother, *Pioneer 10*, had already visited the giant planet a year earlier, but the new kid was determined to go one better. *Pioneer 11* swung by Jupiter, picking up speed from its immense gravity, and flew off to become the first spacecraft to visit Saturn.

It took almost five years to get there, and the science team used this time to debate whether to play it safe and steer *Pioneer 11* around the outer edge of the rings, or try a risky maneuver that would take it right through them. At the time, no one knew how large or dense the ring particles were, and grains even a millimeter across could have destroyed the instruments. Eventually, NASA decided that *Pioneer 11*'s most important job was not to be a daredevil, but to find a path for future spacecraft. It crossed well outside the A ring at a safe distance of 170,000 km (105,600 miles), the same route *Voyager 2* would take two years later.

In addition to gathering the first close-up images of Saturn, *Pioneer 11* also discovered the planet's magnetic field. It found what is now called the F ring and sent back fascinating images of its unusual structure, and it determined that Titan is too cold to support life. Most importantly, though, it opened the door for *Voyager 1*, which arrived in November 1980, and *Voyager 2*, which followed nine months later.

Weeks before its closest encounter, *Voyager 1*'s

The Voyager spacecraft sent back even better images of the F ring, discovered by Pioneer 11 *in 1979.*

◁ *Scientists at the Jet Propulsion Laboratory in Pasadena, California, monitor* Voyager 1's *flyby of Saturn in October 1980.*

superior cameras were already sending back detailed images of the rings, making *Pioneer*'s look like amateur snapshots. *Voyager 1* confirmed that the F ring includes ringlets that are kinked and braided in a way astronomers still cannot explain. It also revealed that the Cassini division is not empty, but simply an area where the particles are less densely packed. The spacecraft did find one section almost devoid of material, however: the precise resonance point with Mimas that Daniel Kirkwood had predicted more than a hundred years before.

Voyager 1 discovered helium in Saturn's atmosphere, measured its wind speeds, discovered three new moons and the G ring, and sent back baffling images of dark bands that cut across the width of the B ring. No one could explain these "spokes," since they seemed to defy Kepler's law: If they were made up of orbiting particles, the outer ones should move more slowly than the inner ones, making a spoke-like shape impossible. (Maxwell and Keeler had made a similar argument when they explained why the rings couldn't be solid.) The spokes are still not completely understood, though they're probably made up of tiny charged particles carried along by Saturn's rotating magnetic field.

Voyager 2 captured even better images of the rings, took Saturn's temperature and measured its density. Both spacecraft also toured the planet's

Both Voyagers *are carrying gold-plated records that include nature sounds, electronic images and greetings in more than 60 languages, in case either spacecraft is ever discovered by other intelligent beings.*

moons and sent back images and data that surpassed anything scientists could have hoped to learn directly from Earth.

Voyager *images revealed mysterious "spokes" in the rings, which appear quickly and change in appearance. They are still not understood by scientists, and the Cassini spacecraft will study these structures more closely when it arrives in 2004.*

What Are You Doing After Work?

After encountering Saturn, *Voyager 1* headed out of the solar system at over 60,000 km (37,300 miles) per hour and is now the most distant man-made object in the universe. *Voyager 2* visited Uranus in 1986 and Neptune in 1989 before speeding toward interstellar space. In 300,000 years, it should be halfway to Sirius, the brightest star in the night sky. Originally designed to last five years, both spacecraft are still transmitting data to Earth, and should do so until about 2020.

Pioneer 11 sent its last communication in November 1995 and is now silently heading toward a visible star in the constellation Aquila. Estimated time of arrival: four million years.

Voyager 2 *during its encounter with the planet Neptune*

Scratching Titan's Surface

Saturn has a large family of moons – at least 18 have been named and other small ones will be added to the list when astronomers confirm their orbits. Unlike Jupiter, though, which has four large moons easily seen through binoculars, the ringed planet has only one heavyweight satellite: Titan.

Named for the giants of Greek mythology, Titan is perhaps the most interesting moon in the solar system. With a diameter of 5,150 km (3,200 miles), it's bigger than Mercury and Pluto, 48 percent larger than Earth's moon, and more than three times the diameter of any other Saturnian satellite. Jupiter's Ganymede barely nudges it out for first place in the solar system. On a clear night, even a small backyard telescope can spot Titan, as Christiaan Huygens was first to do on March 25, 1655.

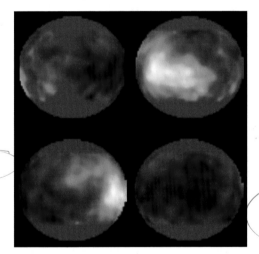

In 1998, the Hubble Space Telescope got these glimpses of Titan's surface. The Cassini spacecraft will use radar to obtain much better images of what lies beneath the clouds.

Other than its exceptional size, though, nothing set Titan apart from other satellites until a 20th-century discovery. In 1907, José Comas Solá was studying Titan from his observatory in Barcelona, Spain, when he noticed a darkening around the moon's edges and suggested it might be an atmosphere. It wasn't until 1944 that Dutch-American astronomer Gerard Kuiper was able prove it with his spectroscope. Kuiper found traces of methane, confirming that Titan is unique – the only moon in the solar system with a significant atmosphere.

When *Voyager 1* flew within 4,000 km (2,500 miles) of the big moon in November 1980, scientists were hoping to get a look at its surface, but a thick orange smog made it invisible. *Voyager* images show Titan as a featureless apricot globe.

◁ *Titan is the only moon in the solar system with a dense atmosphere. The orange haze is produced when sunlight breaks down methane to form other gases.*

Since then, the Hubble Space Telescope has peeked through the haze to reveal something of Titan's surface, though these images do not show much detail. From the way the surface reflects light and radar, it appears to be covered with water ice in some places and darker material in others. At least one region resembles a continent – perhaps a huge mountain. If methane or other hydrocarbons do indeed exist as liquids on the surface, they may have carved valleys and cliffs.

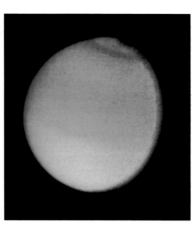

While Voyager *images showed amazing detail on Saturn's other moons, Titan's surface was invisible.*

Voyager 1 did not detect a magnetic field around Titan, which means its interior is not metallic. Most likely, it's about equal parts rock and water ice, perhaps mixed with methane and ammonia, either frozen or in a liquid state.

Though instruments on Earth detected only methane, *Voyager 1* discovered that Titan's atmosphere is actually 90 to 97 percent nitrogen, while methane makes up only 2 to 10 percent. Ultraviolet light from the sun, along with charged particles from Saturn's magnetic field, breaks down the methane to form ethane (C_2H_6) and other hydrocarbons, causing the orange smog that engulfs the moon. What's puzzling is why, after 4.5 billion years, there's still methane in the atmosphere – all of it should have broken down long ago. The only explanation is that Titan must constantly resupply its atmosphere with new methane, but scientists aren't sure how.

Titan's surface pressure is at least 50 percent greater than Earth's, even though the Earth is much larger, and the average temperature is about −184°C (−300°F), similar to that of Saturn itself. No one knows exactly how much weather there is – rain, thunder and lightning are possible, and there are almost certainly winds at high altitudes. Given its dense clouds and great distance from the Sun, it receives only a tiny fraction of the sunlight our planet does, but daytime on Titan would still be about 350 times brighter than a night on Earth with a full moon.

In some ways, Titan's environment resembles Earth's before life evolved. So, is there any possibility of life on Titan? The idea is tantalizing – Titan is one of only half a dozen solar-system bodies where it's at least a remote possibility. But as far as scientists can tell, it's far too cold for liquid water to exist, and without liquid water there can be no life as we know it.

Moon Dance

S aturn's large network of moons is the most orderly in the solar system. Except for the outermost satellites, their orbits are almost perfectly circular and occupy the same plane as Saturn's equator. As they hurtle around, the gravity of each tugs at the others and at the ring particles. Think of it as a celestial dance, with the inner satellites clearing the floor and those at the edge of the rings tapping out the rhythm. Telesto, Tethys and Calypso form a conga line, while Epimetheus and Janus spin their partners round and round. And far off in the distance, Phoebe the rebel marches to her own drummer.

Except for Titan, Saturn's big moons are largely made of frozen water and so are collectively called the icy satellites. And other than Phoebe, most were probably formed around the same time from the same material that makes up the rings. But while they may have a similar makeup, they vary widely in appearance. Indeed, before the *Voyager* spacecraft began sending back their images, no one realized just how many different faces Saturn's moons have.

The four moons closest to Saturn are well within the inner rings. Tiny **Pan** is right inside the Encke division and acts like an orbiting Zamboni,

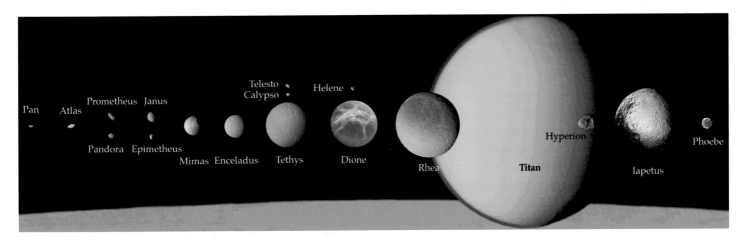

Saturn's 18 major satellites form an impressive lineup. All the moons are to scale in this image, except tiny Pan, Atlas, Telesto, Calypso and Helene, which are shown five times larger.

keeping the gap clear of particles. The slightly larger and irregularly shaped **Atlas** orbits just outside the A ring. Because its gravitational tugs keep this flock of particles organized, Atlas is called the A ring's shepherd satellite.

The next pair are also shepherds. **Prometheus** is just inside the F ring, while **Pandora** circles slightly outside it. Because Prometheus is travelling a little faster than the ring material, its gravity tends to speed up the straggling particles, while Pandora's slower orbit restrains the quick ones. These satellites almost certainly have some relation to the braids and kinks in the F ring, though they don't fully explain these bizarre structures.

In 1966, astronomers reported a new moon orbiting Saturn at about 150,000 km (93,200 miles). Twelve years later, they discovered there were actually two moons sharing almost the exact same orbit, only 50 km (30 miles) apart. **Epimetheus** and **Janus**, which may once have been a single body that was broken up by a collision, are similar in size and play a unique game of orbital leapfrog. About every four years, one passes the other on the inside, nudging the outer moon with its gravity. Instead of colliding, the two satellites then exchange positions and the leader becomes the follower for the next four years.

The rest of Saturn's moons orbit outside the main ring system. **Mimas** was spotted by William Herschel in 1789, but it wasn't until *Voyager 1* that scientists looked this moon in the eye. At least, it looks like an eye – a third of Mimas's diameter is covered by a huge crater. Named for the moon's discoverer, Herschel Crater is 130 km (80 miles) across, with a 6 km (4 mile) mountain in the center. Mimas must be one tough piece of ice to have held together after an impact like that.

Enceladus, also discovered by Herschel, is a little larger than Mimas, but a whole lot brighter. This moon's slick, icy surface reflects almost 100 percent of the sunlight reaching it – if our Moon were covered with the same material it would be 14 times more brilliant. Enceladus has fewer

◁ *An artist's idea of Saturn's shepherd moons, whose gravity and motion maintain the shape of the ring system.*

Mimas's giant crater gives it the appearance of an enormous eyeball. The crater is more than 130 km across, about one-third of the moon's diameter.

While Earth's Moon reflects only about 7 percent of the sunlight that reaches it, Enceladus reflects almost 100 percent.

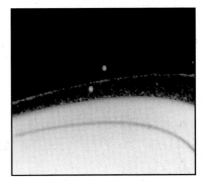

The narrow F ring is bracketed by its two shepherd satellites, Prometheus and Pandora. In this Voyager 2 image, the tiny moons are less than 1,800 kilometers apart.

The Ithaca Chasma, which stretches across the top of Tethys, was probably formed when the surface of the young moon cooled and cracked.

and smaller craters than its neighbors, and parts of its surface appear to have been smoothed over during the past few hundred million years. Scientists aren't sure how this could have happened – ice may have melted and refrozen, but the moon's average temperature is about –200°C (–328°F), so any thawing seems unlikely. Another theory is that the gravity of the nearby moon Dione creates tides that churn up the core of Enceladus, generating heat that could erupt onto the surface. But this seems just as improbable.

Enceladus orbits within the E ring, and another mystery is its relation to the these particles. The E ring has a bluish tinge not seen in the others, so it's likely made of different material. It's also brightest and thickest in the region of Enceladus's orbit, which suggests the moon might have geysers or some other means of spewing out the particles that make up this ring.

Tethys is twice the diameter of Enceladus and is almost entirely water ice. This enormous snowball has two remarkable features: The first is a trench called the Ithaca Chasma, which stretches three-quarters of the way around the satellite. It may be a giant crack that formed billions of years ago when the moon's surface cooled before its interior. The other showpiece is the crater Odysseus – at 400 km (250 miles) across, it's the largest clearly defined crater in the solar system. But unlike Mimas's massive crater, this one rebounded by dozens of kilometers after the impact, suggesting that the collision occurred when the moon was warmer and softer – perhaps partly liquid. Had it been completely solid at the time, it might have been smashed to bits.

Like a queen with her attendants, Tethys is accompanied in its orbit by two almost identical companions. **Telesto** is a tiny, irregularly shaped satellite that orbits 60 degrees in front of Tethys, while **Calypso** is a virtual twin that follows exactly the same distance behind. This tidy choreography is no accident. In order for a small moon to share an orbit with a much bigger one, it must balance the gravity of the large satellite and the planet itself. In 1772, French mathematician Joseph Louis Lagrange discovered that there are only five places where this equilibrium occurs. One of these so-called Langrangian points is 60 degrees in front of the larger body, and another is 60 degrees behind.

Dione has many similarities with Tethys. Both discovered by Cassini, they're almost the same size, though Dione is slightly larger and more dense, which suggests its icy surface surrounds a rocky core. The moon's terrain includes craters, plains and bright, wispy streaks that may have been formed when material erupted from cracks in the surface. Strangely, *Voyager* found that Dione's backwards-facing side has at least as many craters as its leading edge, which is not what one would expect. It's like taking a car ride through the country and finding as many dead bugs on the back window as on the windshield.

During the *Voyager* flybys, scientists noticed that the radio waves emitted by Saturn appear to change (or modulate) every 66 hours – the same time it takes Dione to complete one orbit. Scientists believe these two events are related – Jupiter's moon Io is known to have a similar effect – though they don't understand exactly how.

Like Tethys, Dione is accompanied by a tiny satellite at one of its Langrangian points. **Helene**, only 36 km (22 miles) across, orbits 60 degrees in front of its partner.

Rhea, the second-largest of Saturn's moons, is heavily cratered and has wispy surface features that resemble Dione's. This icy world also harbors a few of its own surprises. While Rhea's surface is covered with craters, *Voyager* images found that the regions near the poles and the equators had smaller pockmarks than the rest of the moon, and they're not sure why.

When **Hyperion** was first spotted by telescopes in 1848, astronomers noticed its orbit is eccentric and its brightness seems to vary. When *Voyager 2* sent back its first images in 1980, it quickly became obvious that this was no ordinary satellite. Unlike every other similarly sized moon in the solar system, Hyperion isn't a sphere but an irregularly shaped boulder that's been likened to a hockey puck and a dented hamburger.

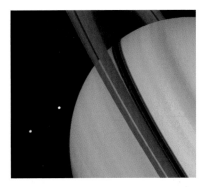

Dione (bottom) and Tethys are almost equal in size.

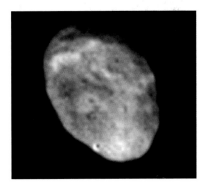

Hyperion is the most oddly shaped large moon in the solar system.

One half of Iapetus reflects 17 times more light than its other half.

Because Phoebe is so far away from Saturn's other satellites, the Voyager spacecrafts returned only blurry images.

Around Saturn and throughout the solar system, gravity locks most moons into what is called a synchronous rotation, which means that the same face always points toward the planet. This is the case with our own Moon – no one saw the far side until a Soviet spacecraft photographed it in 1959. Hyperion, on the other hand, is tumbling head over heels and the speed of its rotation varies unpredictably.

After Huygens discovered Titan in 1655, it remained Saturn's only known satellite for 16 years. Then in October 1671, Cassini peered into his telescope and discovered a second. **Iapetus** is almost three times farther from Saturn than Titan is, and its orbit is tipped 15 degrees from the plane of the rings and the inner moons.

As he observed Iapetus in its 80-day trip around Saturn, Cassini noticed that it regularly grew dimmer and disappeared, only to reappear again about a month later. "It seems," Cassini wrote, "that one part of its surface is not so capable of reflecting the light of the Sun … as the other part is." More than 300 years later, *Voyager* proved him right. One hemisphere of Iapetus has an icy surface that reflects up to 50 percent of the sunlight it receives, while the opposite half is pitch black and reflects as little as 3 percent. The boundary between these two is dramatic, as though the moon were splattered with tar. Some scientists believe the dark coating might be particles cast off by Phoebe – although that tiny moon isn't quite the same color, and it's millions of kilometers away. Other theories suggest the material was created on Iapetus itself, though it remains a mystery.

The outermost satellite is perhaps the oddest of them all. **Phoebe** was the first moon in the solar system to be discovered by photography – William Pickering snapped its picture in 1898. Almost 13 million km (8 million miles) from Saturn – nearly four times farther than Iapetus – it was not seen through a telescope until 1904. During those six intervening years, astronomers noticed something that made it unique.

The other 17 moons, the rings and the planet itself rotate counterclockwise, but Phoebe has a retrograde orbit, meaning it travels in the opposite direction. It could not, then, have formed at the same time as the rest of the Saturn system – it's likely an asteroid that strayed too close to the planet and was captured by its gravity. Phoebe's orbit is also tipped radically compared with the rings and the inner satellites, and it's the only moon with a regular, independent rotation period – that is, it doesn't always show the same face to Saturn.

Plane to See: New Moons of Saturn

Many of Saturn's moons were discovered in bunches – two in 1671-1672, other pairs in 1684, 1789 and 1966, then half a dozen in 1980. In each of these years, Saturn's rings were seen edge-on from Earth, an event known as a ring plane crossing. This occurs every 15 or 16 years, and when it does the rings almost disappear and the satellites pop into view. Until Phoebe was discovered in 1898, all of Saturn's satellites were discovered during ring plane crossings, or just prior to one.

New moons of Saturn are occasionally discovered, but it can take years to confirm them. With the big moons long since charted, any new satellites will be tiny – no more than a few dozen kilometers in diameter – and their orbits are difficult to follow. During the 1995 ring plane crossing, four new moons were tentatively announced, but it was later learned that two were in fact Atlas and Prometheus (which had moved from their predicted locations), and the other two were never seen again.

Then in 2000 astronomers discovered 12 small moons – not in the rings, but 11 million to 23 million km (6.8 million to 14.3 million miles) from Saturn. They range from 8 to 48 km (5 to 30 miles) in diameter and they all have irregular orbits – highly elliptical and sharply tilted in relation to the rings and large moons. Five have retrograde orbits, like Phoebe. The moons seem to be organized in three or four groups, so astronomers believe they may be fragments of larger objects.

In 2003, yet another irregular moon was discovered. When these satellites receive official names from the International Astronomical Union in France, they'll bring the Saturn's total number of known satellites to 31.

Saturn's newest satellite, seen here, was announced in 2003. The tiny moon was discovered by analyzing photographs to detect objects that move against the background of stars.

The Cassini-Huygens Mission

At 4:43 a.m. on October 15, 1997, the air filled with smoke and a thunderous roar broke the morning silence as a rocket lifted off from Cape Canaveral in Florida. Its passenger was *Cassini–Huygens*, the most advanced planetary explorer ever launched, embarking on a 2,450-day flight to Saturn. The two-part spacecraft – the *Cassini* orbiter and the *Huygens* probe – is scheduled to reach Saturn on July 1, 2004. And like its predecessors more than 20 years before, it will change much of what we know about the ringed world.

The *Cassini–Huygens* project is a partnership between the National Aeronautics and Space Administration (NASA), the European Space Agency and the Italian Space

Cassini–Huygens lifts off from Florida's Cape Canaveral on October 15, 1997, on its seven-year journey to Saturn and Titan.

Agency. NASA has primary responsibility for *Cassini*, while *Huygens*'s mission is led by teams in France, Germany, Italy and the United Kingdom as well as the US. In all, 17 countries had a hand in constructing the spacecraft.

While *Pioneer 11* and the *Voyagers* sent back extraordinary data, they were celestial tourists whose close encounters with Saturn lasted mere days. *Cassini*, on the other hand, will remain in the neighbourhood for four years, giving it plenty of time to meet the locals. The *Huygens* probe, meanwhile, will separate from its traveling companion on December 24, 2004, and three weeks later will plunge through Titan's atmosphere to the surface of the giant moon.

◁ The Cassini *spacecraft will orbit Saturn for at least four years, while the small* Huygens *probe will perform its mission in about two and a half hours.*

Getting There is Half the Fun

A rocket to the moon can carry all the fuel it needs to get there and back, but this is impossible for a spacecraft traveling to the outer solar system. If *Cassini* were loaded with enough rocket fuel to get to Saturn, it would be far too heavy to launch. And even if it could get off the ground, the trip would take decades. Spacecraft like *Cassini* use an ingenious technique that saves huge amounts of fuel and dramatically reduces travel time: They get gravity boosts from the planets.

To understand how this works, place a small metal ball on a table and hold a magnet nearby. You'll notice that the closer the ball gets to your magnet, the faster it rolls. Now try pulling the ball along with the force of attraction – being careful not to let the ball actually hit the magnet. "Gravity assists" in space work on the same principle. The spacecraft is accelerated by the planet's gravity, but its navigators time it so the two objects don't collide.

A professional wrestler sometimes hurls himself against the ropes to gain momentum before flying back toward his unfortunate victim. In a similar way, when *Cassini* was launched in October 1997, its headed not outward toward Saturn, but inward at

Venus. The following April, *Cassini* swung by that planet to pick up speed. Eight months later a burst of its rockets turned the spacecraft toward Venus again, where it got another boost that sent it toward Earth. *Cassini* got a third gravity assist from our planet in August 1999 and then headed off toward Jupiter. The giant planet gave the spacecraft its final push on December 30, 2000. These gravity boosts accelerated *Cassini* to a maximum speed of about 72,000 km (45,000 miles) per hour.

Forces always act in both directions so, in theory, when a spacecraft speeds up by borrowing gravity, the planet should slow down. But because of the enormous difference in their sizes, it's like tossing a grain of sand at Mount Everest. The *Voyager* spacecrafts, for example, were accelerated more than 56,000 km (35,000 miles) per hour when they swung by Jupiter, slowing the giant planet down by all of 30 cm (12 inches) per trillion years.

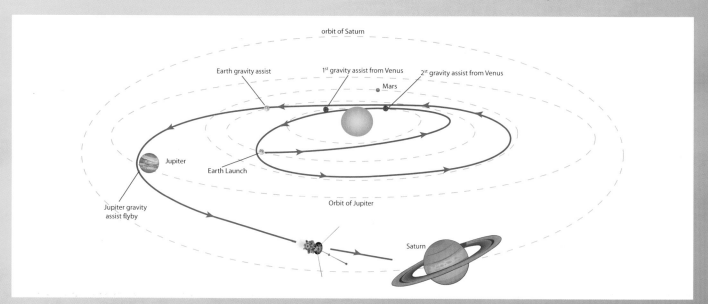

orbit of Saturn

Earth gravity assist   1st gravity assist from Venus   2st gravity assist from Venus

Mars

Jupiter

Earth Launch

Jupiter gravity assist flyby

Orbit of Jupiter

Saturn

The small *Voyager* spacecrafts reached Saturn in just four years, and they whistled by the ringed world at tremendous speeds. Because *Cassini* is so much heavier (about seven times the mass of *Voyager*) its flight will take almost seven years, and because it was designed to go into orbit rather than simply fly past Saturn, it will be traveling much more slowly when it arrives – a mere 56,000 km (35,000 miles) per hour. At Saturn's doorstep, *Cassini* will slow down by firing its thrusters and perform a 90-minute maneuver that will allow the planet's gravity to capture and hold it.

During its four-year mission *Cassini* will make 75 orbits of Saturn, 45 flybys of Titan, and close visits with Phoebe, Iapetus, Hyperion, Rhea, Dione, and (three times) Enceladus. Throughout this time, the spacecraft will stay in touch using the Deep Space Network, a series of three huge antennas spaced about equally on Earth – one in California, another in Spain and the third in Australia. As our planet rotates, at least one of these antennas will always face the spacecraft. But that doesn't mean the science team can control *Cassini* with a joystick, as though it were a spaceship in a video game. Because of the great distance, it will take 68 to 84 minutes for radio signals to reach Earth from Saturn, and the same time to send them back. Since a delay of two or three hours could be fatal, *Cassini* has been programmed to detect and solve many problems on its own. If necessary, it

The Huygens *probe and parts of the* Cassini *spacecraft were wrapped in thermal blankets to protect them from the frigid temperatures of outer space.*

can even stop all of its scientific work and go into a safe hibernation until the crisis is addressed by ground control.

As *Cassini* hurtles through space, it's subjected to extremes of heat and cold. Most of the spacecraft is wrapped like a baked potato in shiny thermal blankets, but until it was about 400 million km (250 million miles) from the Sun, it

Cassini is 6.8 meters high and weighs more than 5,600 kg, making it the largest planetary spacecraft ever launched.

also throw the instruments off, so several are kept warm by heaters.

Of course, scientific instruments and heaters need electricity. While Earth-orbiting satellites – and even missions to Mars – normally use solar energy, Saturn is too far from the Sun to make this possible. Instead, Cassini gets its juice from three RTGs – radioisotope thermoelectric generators. Because these nuclear power sources contain over 32 kg (70 pounds) of radioactive plutonium, some people worried that the spacecraft would pose a danger had it exploded during launch, or when it flew by Earth. The risks were extremely small – the only way this substance can harm humans is if it's inhaled and lodges in a person's lungs for years. Nonetheless, all spacecraft carrying RTGs must go thorough rigorous safety tests, and Cassini passed all of them comfortably.

Repair crews can't get to Saturn, so the spacecraft has as few moving parts as possible. For example, while the Voyagers used tape decks to store information, Cassini uses solid-state recorders – a bit like replacing floppy disks with RAM chips in a home computer. They will store the information collected by the instruments and transmit it when the main antenna is pointed at Earth. In a typical day, Cassini will spend about 15 hours gathering data and 9 hours beaming it home.

could still have been damaged by solar rays. To prevent this, Cassini used its big antenna dish like a beach umbrella to shade the sensitive science instruments. Chilling temperatures in space can

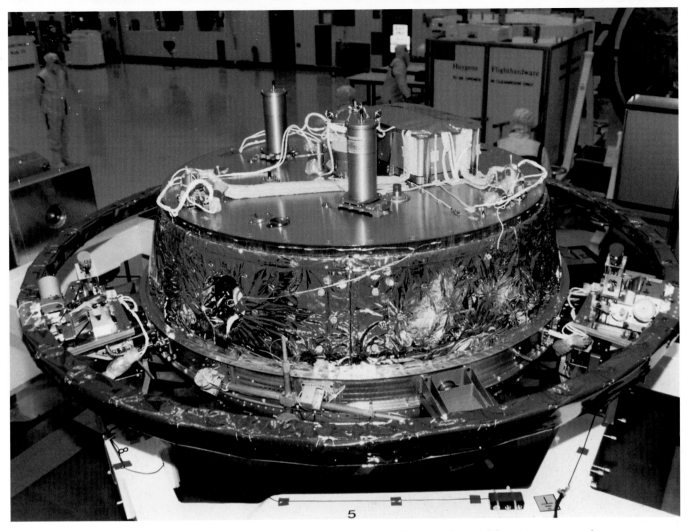

5

Because the Huygens probe will enter Titan's atmosphere traveling more than six kilometers per second, its instruments are protected by a heat shield.

If *Cassini* is a marathoner, *Huygens* is like a sprinter who spends thousands of hours training for a race that's over in 10 seconds. When the probe arrives at its destination after journeying some 3 billion kilometers, it will perform its mission in Titan's atmosphere in just two and a half hours.

On Christmas Eve, 2004, *Huygens* will bid farewell to *Cassini* after almost seven years. Small explosions will pop open the spring-loaded mechanisms that hold the two spacecraft together, and the probe will be pushed away at about 30 cm (12 inches) per second. A gentle spin of seven revolutions per minute will help keep it steady.

Huygens has no rocket thrusters of its own, so it will simply coast to Titan. It will sleep during the 21-day journey, taking no measurements or photographs. Its batteries will run only an alarm clock, set to rouse the probe 28 minutes before it enters Titan's atmosphere on January 14, 2005.

The probe will enter the atmosphere traveling over 6 kilometers (3.7 miles) per second, creating enough friction to burn it to a crisp if it were not protected. Therefore, *Huygens's* front end is fitted with a shield that looks like an upside-down mushroom cap. It's covered with special tiles that can withstand the tremendous heat.

Once the probe hits the atmosphere, friction will slow it to about 400 meters (437 yards) per second. Then a series of three parachutes goes to work. First, a pilot chute will pull off the probe's back cover, then the much larger main chute will slow the probe down and make the front heat shield unnecessary. Once it comes off, *Huygens* is open for business and the scientific measurements will begin. After 15 minutes, the main parachute will be released and a smaller one, called a drogue chute, will take *Huygens* all the way to the surface.

The descent will take 120 to 150 minutes, during which six instruments will collect first-hand information about Titan's atmosphere and take the first photographs of the moon's surface. A specific landing site was chosen on the day side of Titan – that is, facing the Sun – but no one really knows what the surface will be like there. If the probe does survive the thud (or splash, as the case may be) it will continue gathering information for up to two hours, until its batteries die or the computers become too cold to function.

Huygens will relay its data to *Cassini*, which will in turn send it to Earth. Originally, the science team planned for *Cassini* to be just 1,200 km (745 miles) above Titan during the probe's descent. But in February 2000, they discovered a problem with the probe's communication system and had to change the schedule. *Cassini* will now be more than 60,000 km (37,285 miles) away during the probe's descent, but this extra distance should not pose any problems – the orbiter will still be able to listen to *Huygens* for more than four hours before it dips below the horizon. None of the scientific goals had to be changed because of the glitch – the only significant cost was the extra fuel *Cassini* used to change its course.

Before astronauts visited the Moon, several spacecrafts landed there to see what the surface would be like. ▷
By comparison, scientists do not even know whether Huygens will touch down on land or in an ocean.

Cassini's Scientific Goals

Around Saturn:

- measure the amounts of different gases in the atmosphere
- measure winds, temperature and cloud features
- look for lightning in the atmosphere
- gather data about the planet's interior structure and how it formed
- observe how charged particles around Saturn react with its magnetic field

In the rings:

- get a better understanding of the puzzling structures revealed by *Voyagers 1* and *2*
- investigate interactions between the rings and satellites
- study the material and distribution of the particles
- analyze the dust and meteoroids near the rings

At Titan:

- measure the amounts of different gases in the atmosphere
- measure winds, temperature and cloud features
- determine if any liquid is present on the surface
- map solid surface features and determine their composition
- study how Titan interacts with the solar wind and with Saturn's magnetosphere

At the other moons:

- investigate the causes of surface features, such as the dark material on Iapetus and the wisps on Dione, and detect any geological activity, such as geysers on Enceladus
- detect any thin atmospheres that cannot be measured from Earth
- determine whether the satellites are the source of the ring particles
- learn how the satellites interact with the magnetosphere and rings
- look for new moons within the ring system and beyond

Cassini's Scientific Instruments

Imaging Science Subsystem. This pair of digital cameras – one for close-ups, the other for wide angles – will photograph the planet, rings and satellites, as well the atmospheres of Saturn and Titan.

Visible and Infrared Mapping Spectrometer, Ultraviolet Imaging Spectrograph and **Composite Infrared Spectrometer**. These instruments measure the way different wavelengths of light are absorbed, reflected or emitted by gases (such as those in an atmosphere) and solids (such as ring particles or a moon's surface). Because every substance does this differently, scientists can identify the material and measure its temperature.

Cassini Radar. This radar will emit signals designed to cut through the hazy atmosphere and map Titan's surface.

Radio Science Subsystem. In tandem with huge antennas on Earth, this device will send radio waves through the rings to learn more about their particle size and structure, and through the atmospheres of Saturn and Titan to measure their composition, pressure and temperature. It will also estimate the masses of the planet and its satellites.

Dual Technique Magnetometer. Because the electric currents and magnetic components of other instruments would throw off its readings, the magnetometer is located on an 11-meter boom that keeps it well away from the rest of the spacecraft. From there it will analyze Saturn's magnetic field, determine whether Titan has one at all, and measure many other processes in the magnetosphere.

Cassini Plasma Spectrometer, Ion and Neutral Mass Spectrometer and **Magnetospheric Imaging Instrument**. This trio of instruments, mounted together, will measure the particles (both charged and neutral) in Saturn's magnetosphere and Titan's upper atmosphere. They will also observe how these particles interact with the solar wind, the rings and satellites, and with each other.

Radio and Plasma Wave Science Instrument. One of several sensors that will gather information about the particles and radiation in Saturn's magnetosphere, this device will also analyze lightning in the atmosphere and study how the icy satellites and rings interact.

Cosmic Dust Analyzer. This bucket-shaped instrument will scoop up tiny particles of dust and ice and measure their size, density and charge. It will analyze the particles in the mysterious E ring and elsewhere to determine whether Saturn's satellites are the source of this material.

Huygens's Scientific Goals

- determine the amounts of different gases in Titan's atmosphere, including complex molecules that may form when nitrogen and methane (which is made of carbon and hydrogen) break down
- investigate the energy sources – including the Sun, Saturn's magnetosphere and cosmic rays – that cause chemical reactions in the atmosphere
- study the properties of the smog and clouds
- measure winds and temperatures
- obtain the first detailed look at the surface and gather information to help determine the internal structure
- investigate the upper atmosphere and the charged particles that exist there

Huygens's Scientific Instruments

Huygens Atmospheric Structure Instrument. These sensors will determine the density of the atmosphere by measuring the rate at which the probe slows down as it descends. They will also measure wind speed, pressure, electrical activity, temperature and, if the probe lands in a liquid, wave motions. This instrument includes a microphone that will listen for thunder and rain.

Doppler Wind Experiment. This series of measurements will precisely determine wind direction and speeds (to within one meter per second) at different altitudes. The instruments will also help record the probe's movements.

Descent Imager and Spectral Radiometer. This includes the probe's cameras, which will photograph the surface and the clouds. Sensors will determine the density of the smog particles by measuring how much they block out sunlight. A lamp will also switch on before landing and a spectrometer will measure how the surface absorbs light, indicating what it's made of.

Gas Chromatograph Mass Spectrometer. This instrument will analyze the gases collected from the atmosphere and determine their molecular mass. If the probe lands safely, it will also examine the surface material.

Aerosol Collector and Pyrolyser. This device will collect samples of particles in the atmosphere at two different altitudes. In both cases it will heat the particles in an oven to vaporize them, and then send them to the Gas Chromatograph Mass Spectrometer to be analyzed.

Surface Science Package. These sensors will determine what the surface is made of by using a number of simple measurements. They will examine how the surface absorbs light, and gauge its hardness by how abruptly the probe slows down when it lands. If the surface is liquid, they will measure the depth and analyze a sample.

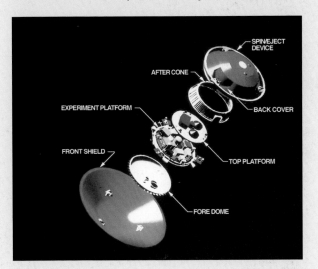

The Huygens probe – shown here with its components separated – will carry six sensitive instruments into the atmosphere of Titan. Other electronic equipment will remain attached to Cassini in order to track the probe and receive the data it transmits.

Observing Saturn

Ask amateur astronomers what turned them on to the hobby and many will recall their first glimpse of Saturn through a backyard telescope. It's not the brightest planet, nor the biggest in the eyepiece, but only Saturn is three-dimensional. Even a small scope with 30x magnification will reveal its unique shape, and on a clear night you can even glimpse Titan. Bigger telescopes clearly show the Cassini division, several other moons and some details on the surface.

If you don't own a telescope, you can still follow Saturn's jaunt across the sky. The planet is visible for about nine months a year – during the others it's above the horizon only during daylight. Within this window of visibility, however, Saturn is sometimes only prominent in the wee hours, when all but the most ardent stargazers are in bed.

The best time to look for Saturn is when it's at opposition. This occurs when the Earth lies directly between Saturn and the Sun – the same arrangement that makes the Moon appear full. For a few weeks before and after opposition – which occurs every 378 days – the planet is at its brightest and remains high in the sky all night long, making it easy to spot if you know where to look.

During its slow orbit, Saturn spends several months to more than two years in each of the 12 zodiac constellations. Until mid-2005 it's in Gemini, near the two bright stars Castor and Pollux. Then it moves into Cancer until late in 2006. In relation to the stars, Saturn normally moves a little bit east each night, but for several weeks before opposition it appears to move backwards as the Earth approaches and then passes Saturn in its orbit. Don't worry if you're confused – this motion puzzled astronomers for centuries, after all. Astronomy magazines, websites and software can help you track down the planet's exact location.

If you were to watch Saturn through a telescope over several years, you'd see the changing face that baffled Galileo in the 1600s. The rings were opened their widest in 2002 and will slowly close until they seem to disappear during the next ring plane crossing in 2009.

By that time, *Cassini* will have finished its main mission but, if all goes well, it will still be in orbit around Saturn, putting in overtime. And as scientists study the information *Cassini* sends back, the ringed planet may just seem a little closer than once it did.

Year	Ring Tilt	Year	Ring Tilt
2003 January–May, August–September in the constellation Gemini		2007 January–July, September–December in the constellation Leo	
2004 January–May, August–September in the constellation Gemini		2008 January–July, September–December in the constellation Leo	
2005 January–June, September–December in the constellation Gemini, then Cancer		2009 January–August, October–December in the constellation Leo, then Virgo	
2006 January–June, September–December in the constellation Cancer, then Leo		2010 January–August, November–December in the constellation Virgo	

The Ring System

Ring feature	Distance from Saturn (km)	Width (km)	Description	Discovery
D ring	67,000	7,500	Faint, diffuse ringlets made up of tiny particles at the top of Saturn's atmosphere.	Pierre Guerin, 1969, confirmed by *Pioneer 11*, 1979.
C ring	74,500	17,500	Called the "crepe ring," because it is semi-transparent. Visible in a telescope under perfect conditions.	Seen as early as 1664; identified by William and George Bond, 1850.
B ring	92,500	25,500	The brightest and densest band, it contains much of the ring system's total mass.	Before 1675, seen as the brightest part of a single ring.
Cassini division	118,000	4,800	Once thought to be empty, it contains several ringlets and many particles.	Giovanni Cassini, 1675.

Ring feature	Distance from Saturn (km)	Width (km)	Description	Discovery
A ring	122,000	14,500	With sharply defined edges and gaps, it's the second-brightest ring in the system.	Before 1675, seen as the dimmer part of a single ring.
Encke division	133,500	325	Almost devoid of particles, probably because it's been cleared by Pan, which orbits inside it.	Confirmed by James Keeler, 1888, though named for Johann Encke.
F ring	140,000	variable (about 50)	Complex kinked and braided strands that are non-circular.	*Pioneer 11*, 1979.
G ring	170,000	variable (about 8,000)	Ill-defined, tenuous ring that contains the orbits of Epimetheus and Janus.	*Voyager 1*, 1980.
E ring	180,000	up to 300,000	Bluish micron-sized particles that may be supplied by Enceladus.	Walter Feibelman, 1967.

The Satellites

Moon	Distance from Saturn (km)	Maximum diameter (km)	Orbital period (days)	Relationship with rings and other moons	Discovery
Pan	133,583	20	0.56	Inside Encke division, which it clears of particles.	Mark Showalter, 1990.
Atlas	137,640	37	0.60	Shepherds outer edge of A ring.	Richard Terrile and *Voyager 1*, 1980.
Prometheus	139,350	145	0.61	Inner shepherd for F ring.	Andrew Collins and *Voyager 1*, 1980.
Pandora	141,700	110	0.63	Outer shepherd for F ring.	Andrew Collins and *Voyager 1*, 1980.
Epimetheus	151,442	144	0.69	Exchanges positions with Janus every four years.	Richard Walker, 1966, confirmed by Stephen Larson and John Fountain, 1978.
Janus	151,472	194	0.69	Exchanges positions with Epimetheus every four years.	Audouin Dollfus, 1966.
Mimas	185,520	418	0.94	Its orbital resonance keeps Cassini division clear.	William Herschel, 1789.
Enceladus	238,020	512	1.37	May be the source of bluish material in E ring.	William Herschel, 1789.

Moon	Distance from Saturn (km)	Maximum diameter (km)	Orbital period (days)	Relationship with rings and other moons	Discovery
Telesto	294,660	30	1.89	Orbits at Lagrangian point in front of Tethys.	Bradford Smith and others, 1980.
Tethys	294,660	1,071	1.89	Orbits directly between Calypso and Tethys.	Giovanni Cassini, 1684.
Calypso	294,660	30	1.89	Orbits at Lagrangian point behind Tethys.	Dan Pascu and others, 1980.
Helene	377,400	35	2.74	Orbits at Lagrangian point in front of Dione.	Pierre Laques and Jean Lecacheux, 1980.
Dione	377,400	1,120	2.74	Shares orbit with Helene.	Giovanni Cassini, 1684.
Rhea	527,040	1,528	4.52	Second in size only to Titan.	Giovanni Cassini, 1672.
Titan	1,221,850	5,150	15.95	Only moon with an atmosphere.	Christiaan Huygens, 1655.
Hyperion	1,481,100	360	21.28	Largest irregularly shaped moon in the solar system.	William Bond, George Bond and William Lassell, 1848.
Iapetus	3,561,300	1,436	79.33	May sweep up dark material from Phoebe.	Giovanni Cassini, 1671.
Phoebe	12,952,000	230	550.48	Orbits in the opposite direction to all other large moons of Saturn.	William Pickering, 1898.

Saturn Versus Earth

Property	Saturn	Earth	Saturn:Earth ratio
Mass	568.46×10^{24} kg	5.87×10^{24} kg	95.16
Volume	$82{,}713 \times 10^{10}$ km³	108.32×10^{10} km³	763.6
Diameter at equator	120,536 km	12,756 km	9.45
Density	0.69 g/cm³	5.52 g/cm³	0.13
Acceleration of gravity	8.96 m/s²	9.78 m/s²	0.92
Average distance from sun	1,429,400,000 km	149,598,000 km	9.55
Orbital period	29.46 years	365.25 days	29.46
Rotation period	10.66 hours	23.93 hours	0.45
Average orbital speed	9.64 m/s	29.79 m/s	0.33
Tilt of axis	26.7°	23.5°	1.14
Tilt of orbit relative to ecliptic	2.5°	0	–
Average temperature	88 K (–185°C)	288 K (15°C)	–

More of the Rings

To learn more about Saturn and the *Cassini–Huygens* mission, visit these websites:

Cassini–Huygens Mission
saturn.jpl.nasa.gov
Track the progress of *Cassini–Huygens* throughout the mission at its official home page. Includes complete information about the planet and spacecraft, a special section for kids, and resources for teachers, including the Saturn Educator Guide.

European Space Agency: Huygens
sci.esa.int/huygens
Details about the European-built *Huygens* probe and its mission.

Solar System Exploration
sse.jpl.nasa.gov
This NASA site presents an overview of all the planets and their moons and explains the spacecraft missions that have explored them.

Voyager
www.jpl.nasa.gov/voyager
A rich archive of images and information about the *Voyager* missions, including their many discoveries at Saturn.

The Nine Planets
www.nineplanets.org
An overview of what we know about the solar system, including all the planets and their satellites.

Views of the Solar System
www.solarviews.com
A thorough introduction to the planets, including the history of naked-eye and spacecraft discoveries, plus images and animations.

Astronomy for Kids
www.dustbunny.com/afk
Includes regularly updated charts to help observers find Saturn and other celestial objects in the night sky.

Author's Note

This book is for Jaimie and Erick.
I'm particularly thankful for the assistance of Ellis Miner of the Jet Propulsion Laboratory in Pasadena, California; Jean-Pierre Lebreton of the European Space Agency in Noordwijk, the Netherlands; and Philip Nicholson of Cornell University in Ithaca, New York, who not only interrupted their busy schedules, but also shared their enthusiasm, which I hope comes through in this book.

Thanks also to Tom Bolton of the University of Toronto, Amanda Bosh at the Lowell Observatory, Brett Gladman of the University of British Columbia, JJ Kavelaars of McMaster University, Hirini Melbourne of the University of Waikato (New Zealand), Monica Talevi at ESA, and David Morrison and Alice Wessen at NASA.